STRIKING DISTANCE

By the same author:

Broken Moon (Oxford Poets, 1987)
Changing the Subject (Oxford Poets, 1990)

STRIKING DISTANCE

Carole Satyamurti

Oxford New York
OXFORD UNIVERSITY PRESS
1994

Oxford University Press, Walton Street, Oxford OX2 6DP
Oxford New York Toronto
Delhi Bombay Calcutta Madras Karachi
Kuala Lumpur Singapore Hong Kong Tokyo
Nairobi Dar es Salaam Cape Town
Melbourne Auckland Madrid
and associated companies in
Berlin Ibadan

Oxford is a trade mark of Oxford University Press

First published in Oxford Poets
as an Oxford University Press paperback 1994

British Library Cataloguing in Publication Data
Data available

Library of Congress Cataloging in Publication Data
Satyamurti, Carole
Striking distance / Carole Satyamurti.
p. cm.—(Oxford poets)
I. Title, II. Series.
PR6069.A776S77 1994 821'.914—dc20 94-9235
ISBN 0-19-282386-8

1 3 5 7 9 10 8 6 4 2

Typeset by Rowland Phototypesetting Ltd, Bury St Edmunds, Suffolk
Printed in Hong Kong

For Emma, and Martin

ACKNOWLEDGEMENTS

Acknowledgements are due to the editors of the following publications in which some of these poems first appeared: *Critical Quarterly*, *Independent on Sunday*, *Spectator*, *London Magazine*, *Illuminations*, *Acumen*, *Poetry Review*, *The Times Literary Supplement*, *Iron*, *Poetry Book Society Anthology 2*, ed. Anne Stevenson (Hutchinson 1991), *The Blue Nose Anthology* (Blue Nose Press 1993), *Arvon International Poetry Competition Anthology* (Arvon Foundation 1989), *Klaonica: Poems for Bosnia*, ed. Ken Smith and Judi Benson (Bloodaxe 1993), *The Forward Book of Poetry 1994*.

'There Will Come a Time' was commissioned as part of the 1992 Poetry International on the South Bank; 'Tuesday at the Office' by the Scunthorpe Musical Festival. 'Sister Ship' won a Duncan Lawrie prize in the 1989 Arvon International Poetry Competition.

The author also acknowledges the permission of Farrar, Straus & Giroux, Inc. to reprint as epigraph the excerpt from *The Complete Poems 1927–1979* by Elizabeth Bishop. Copyright © 1979, 1983 by Alice Helen Methfessel.

Why should I be my aunt
or me, or anyone?

—Elizabeth Bishop,
'In the Waiting Room'

CONTENTS

I Skin Distance

This Morning 11
Waiting Room 12
The Fall 13
The Derek Game 14
Passed On 15
Coat for an Undergraduate 16
File Past 17
You Make Your Bed 18
Skin Distance 19
Il Conto 20
Tide, Turning 21
Gifts 22
Woman in Brown 23
Presents for Duncan 24
Death Speaks After the Tone 25
Because It's There 26
The Way We Live Now 27
Where Are You? 28
Moment 29
The Present 30
One 31
Diabolus in Musica 32
Our Peacock 33

II Striking Distance

Out of Reach 37
Crossing the Border 38
Striking Distance 40
Thanatos 42
There Will Come a Time 43
Advent in Bratislava 44
The Trial of Lyman Atkins 45
America 46

SISTER SHIP 49
1 Relatives
2 Divers
3 Afterwards
4 Sister Ship
5 Not Us
6 Cafeteria
7 Anniversary
This England 54
Turning Point 55
Cutting Loose 56
Tuesday at the Office 57
Fatima 58
Blurred Vision 59
Life in Tall Houses 60
Wrong-Footed 61
Ourstory 62
The Smell of Sweat 63

I

Skin Distance

THIS MORNING

Creation might have been like this,
early sun stencilling the leaves
of the first ever walnut trees,
and the cows beside them splashes
of caramel, coffee, apricot, vanilla,
drifting as if under water
in breeze-fractured light.

I have the eyes of the academy,
mince the natural world into word-burgers
seasoned with disappointment.
These lovely prelapsarian cows
are a poem, generous in conception,
perfectly achieved, rhythms,
rhyming, untranslatable.

My gaze makes them alien to themselves.
Bashful, they shift their stout elegance,
breathe soft, uneasy huffs, wrestling
with doubt. But as the church clock strikes
seven, twice over, we are between times,
and simply the world's new inhabitants
staring at the other, staring.

WAITING ROOM

It was slippery blind surfaces,
rushing waterways, a distant drum.

It was drifting at anchor, warm
power of stretch and kick.

Sometimes, there was red to dance to,
and songs: mother-sounds without edges.

I'd answer, but nothing came.
My lips were always practising.

A mouth-fit thumb or a toe
would come to console me.

A time of loud words bruising
against each other; then a giddy shock

hurled me against the dim-lit screen,
unharmed, but understanding I didn't have

the temperament for silent suffering,
that this was the moment to

take on gravity, to haul myself
round, and out on a tide of cries.

THE FALL

The rest of your life starts
when a world of snug non-sense
you've not imagined could be otherwise
turns mean, and there you are,
the usual you and getting
smacked for it, not understanding why.

Did you dream rusks, pat-a-cake, the bliss
you summoned when you squeezed your eyes?
Now smiles are weather.
You learn rain-dance and ritual,
slant looks, pent farts,
the cussedness of spoon and fork.

So much forbidden, you never know all
the names for it. You punish your dolls
for their mistakes, and feel quite cheerful;
only sometimes there's a pellet in your mouth
you can't spit out or swallow,
the bitterness of crusts you're stuck with.

THE DEREK GAME

The flavour of our mothers' milk was
moderation. Bites must be not too big;
laughter, colours, not too loud.
A life of tight plaits, garters, kirby-grips;
and bristly necks, for boys.

Neighbours were much like us,
visible tips smiling the time of day.
Herbaceous borders stood correct
as cold front rooms, where surly pianos
dozed, dreaming Ivor Novello.

But then there was Derek, 'not quite right',
hair wild, saliva sparkling,
who danced unasked up every garden path,
talked to anyone, joyfully loud-hailing
the coal man, cats, our laburnum tree.

We ran, hid, purple with held breath;
perhaps we thought it was infectious.
Sometimes, behind the shed, we'd play
the Derek game, scaring each other,
feeling our way inside extravagance.

PASSED ON

Before, this box contained my mother.
For months she'd sent me out for index cards,
scribbled with a squirrel concentration
while I'd nag at her, seeing strength
drain, ink-blue, from her finger-ends
providing for a string of hard winters
I was trying not to understand.

Only after, opening it, I saw
how she'd rendered herself down from flesh
to paper, alphabetical; there for me
in every way she could anticipate
  —Acupuncture: conditions suited to
  —Books to read by age twenty-one
  —Choux pastry: how to make, when to use.

The cards looked after me. I'd shuffle them
to almost hear her speak. Then, my days
were box-shaped (or was I playing safe?)
for every doubt or choice, a card that fitted
  —Exams: the best revision strategy
  —Flowers: cut, how to make them last
  —Greece: the men, what you need to know.

But then they seemed to shrink. I'd turn them over,
find them blank; the edges furred, mute,
whole areas wrong, or missing. Had she known?
The language pointed to what wasn't said.
I'd add notes of my own, strange beside
her urgent dogmatism, loosening grip
  –infinitives never telling love
  lust single issue politics when
  don't hopeless careful trust.

On the beach, I built a hollow cairn,
tipped in the cards. Then I let her go.
The smoke rose thin and clear, slowly blurred.
I've kept the box for diaries, like this.

COAT FOR AN UNDERGRADUATE

From Italy, by way of Harrods.
I snatched it from a wealthy tourist,
this perfect coat, size 9–10 years,
wool and cashmere, silk, exquisite detail.
It flared as I revolved it on its hanger,
deep folds embracing light.

We're excited by it—the cut, the cost.
Aren't we both imagining
it will make you, too, perfect,
propel you down the Broad firm-footed,
give you stature,
free all stiffnesses?

On you, the hem dips,
shoulders poke, empty,
top button, unmanageable.
I rage as I pick open seams,
pin, tack, cut, cack-handed,
compromise the marvellous finish.
Sometimes your whole life seems
make do and never mended enough.

Look—in the country of the possible,
this is a real transformation.
Child-sized you may be, but here is
your come-of-age reflection;
and you're thinking tall
as you drive away in your red car,
green boots, and the perfect enough coat,
swinging, sock-it-to-them, classy blue.

FILE PAST

While his back's turned I slip inside
my pink, fat file,
the cover flopping shut behind me.
It's hot, airless.

I start to chew through layers of forms,
letters, case conference reports,
leaving a hole the size of me.
I get thirsty.

I notice names. Muscles twitch,
remembering—eyelids, anus, fists.
Such a little hand. Put it here.
Love. Aah. Mustn't tell or. Love.

I bite through love
and all the other vomit words
—Security, Care, Sharing, Come to terms—
I can only say with a funny accent.

My name is scattered everywhere
but I'm not in this wad of bits,
postcard from Hayling Island (Dear Mis Jons)
this People In My World diagram,

Life Story Book—my life, their story.
I burrow past all those women
I called Aunty, Mum, Mummy Sue,
twisting fingers. They didn't notice.

And here is a girl who glares with my eyes;
weird clothes, greasy hair, my sort of age
—my mother there are almost no names left for.
Ref. file CP/62/103.

YOU MAKE YOUR BED

You make your bed, precisely not to lie on it
but to confine the disorder of the night
in mitred corners, unruffled surfaces.

Morning's already clamouring in your head.
This is a stand against incompetence,
at least one perfectly accomplished act.

All day it's what you've put behind you,
an infant place that, with the dark, draws you
down, and back; an enticing book

whose soft covers you open, slipping into
those quotidian rehearsals—love
and sleep. And dream irresponsibly

until the bell clangs for the next round.
Loth to climb out, you know you should be glad
you can. You make your bed precisely.

SKIN DISTANCE

Can you imagine this?

You're sitting on the tube
opposite a strapping lad, black,
late teens, seventeen stone-ish.
You haven't noticed him particularly
until he fetches out a jar of
baby-food (creamed carrots),
levers the lid off with his teeth,
spoons up large mouthfuls.

Urgently now, he follows
with turkey dinner, banana delight
sunshine breakfast, rice and pears . . .

And you're feeling queasy, seeing
who you are is accidental
and there's only a couple of skins
between being you, and tipping over
into the life of a young, black, male,
seventeen-stone baby-food junkie
who doesn't seem to care who notices
or what they're thinking.

Can you feel the slither of semolina
creeping across your tongue,
your limbs becoming heavier?

IL CONTO

for Michael Blumenthal

You pay for the fifteenth-century palazzo.
You pay for three smiles in evening dress,
for just the right degree of deference.

You pay for the dexterity of wrists,
the flattering chorale of glass and silver;
for the proffered chair, flourished napery,

the snowy acres of the sommelier's waistcoat.
You pay for the plush hush of the ladies' room,
tiny tablets of geranium soap.

Pasta that leaves a coating on your teeth
is not, you tell yourself, the point. You pay
for the precision sprinkling of *parmigiano*,

for the slick ripple of Gershwin; the magnetism
of Fame and Money at the other tables;
for candle-flame that gilds you, draws you in.

You pay for your reflection in the window
where Dukes' mistresses once leaned. You pay
for a starring role in your own Sweet Life.

And you pay for the centuries-old tact
with which you're not pressed to a second course,
with which the *portiere* averts his eyes.

TIDE, TURNING

Sliced flat, androgynous,
a child again with all to play for,
I've come here to encounter the sea.

Clownish, she catches me out, slap on the back
no joke—though I laugh
spluttering while she collapses in ripples.

Mother, she rocks me on her slack, dark breast.
She's quickly bored,
spits in my eye, fiddles with my hair.

Lover with a wicked past, her licks
and soft caresses turn to rough trade
throwing me off balance.

I am a fish she'll try to suck
down to her pebble bed; I am a rod
erect as she curls above me.

Too quick for her—I slice across
her oily invitation, and up
through shards of savage light.

She cuffs me to size
with her grey, glass paw
reminding me who was here first

—and will be here, long after
anyone who reckons she's paid off
her debt to the moon.

GIFTS

When I think of her, I see that swift
flick of her hair,
how she'd stroke it, brush it,
lowering her head to where the light
sought out the shine;
she called it her gift.

There was a way she lived in music, rare,
like her last illness.
Yet keyboard brilliance
was unremarkable to her, as I
might think of speaking;
or of hair.

Perhaps because by then she had begun
to see it fall
to medical heroics,
she loved it more than gifts she'd only lose
when étude, fugue and hair
would be all one.

Or maybe, in the mirror, her eyes met
the aureole,
brighter than she could bear,
of an enormity—so that she fastened
onto her hair's lustre
all value, all regret.

WOMAN IN BROWN

Woman in brown
almost not there at all
a shadow, shadowless
propped into a chair against the wall,
eyes, semi-breves,
wrists hanging slack:
anaglyptic undersides of leaves.

Imagine her against light—there,
where sun is shafting through the door.
You'd see through skin
to peristalsis, respiration,
the automatic busyness of circulation
—a clock in a deserted house;
action there's no longer reason for.

But for a chance ordering of matter
this woman might have been
me. But she is the one in brown
and I can wrap her up in verse,
drive back to town
as her thoughts leak gently out
to blizzard on the television screen.

PRESENTS FOR DUNCAN

for Nancy Stepan

Ten years ago, I could have brought a book
of Gramsci's, say, or Larkin's. You'd have asked
my views on Europe, taken me to look
at a new camellia; maybe we'd have driven
somewhere for a concert—Bach or Bruch.

Even last year, there might have been a walk,
you in a wheelchair, through the ordered grounds
to the pub beyond the gate. Then, you could talk
about the people in old photographs,
and tell a kestrel from a sparrow hawk.

Opportune, that in middle age we're blind
to future selves, will not imagine some
arterial sabotage could make a mind
that rocketed shrink to random sparks
(since entropy's impartially unkind).

Today, as you catch sight of me, you scream
and wail, knowing yourself, in that moment,
lost. Then reaching, parrot-slow, you seem
content to cram your mouth with an old sweetness:
all the world in a chocolate orange cream.

DEATH SPEAKS AFTER THE TONE

What shall I say to you,
your uni-person voice designed to suit
sales-people, colleagues, closest friends, your son?
You must see it won't do

for me, who know you best.
I was there before you knew yourself.
I understand the fluxion of your heart,
its innermost recess.

'I'm not here right now'
—technology leads you to murder sense.
Or is it fear makes you indifferent
to what words say, and how?

If you'd allow me in,
let me speak directly to you, soon
you'd see I am essential to your life,
its spiritual twin.

You think you'll put me off
if you cower behind that amputated voice
there, not there, never there for me.
But I've time enough.

I'll never go away.
I'll be here, pressed against your grille,
smoked glass, net curtain of a voice,
day after day after day.

BECAUSE IT'S THERE

for Judith Ravenscroft

After you've told yourself
you don't have to go in unless you want
and after a voice whispers *wimp*;

after you've waited
for the nauseous bladderwrack to set up
ambush elsewhere;

after you've not undressed
in case that blonde man by the breakwater
is the M6 murderer;

after you've reflected
on the effluent of Criccieth and Llanystumdwy,
visualized a million little mouths;

you arrive at a numbness of the spirit
where nothing is left but to strip off,
pretending your body's someone else's, and

enter the impossible cold.
And the sea rewards you
with several salty smackers, asking

why the fuss? Directing your gaze
to where the sun flings down a challenge
over the white caps of Snowdonia.

THE WAY WE LIVE NOW

I'm walking into *La Porchetta*
with Daughter and ex-Husband
when I spot ex-Lover and new woman
sharing a tricolore salad, mozzarella
soaring mouthwards on a single fork.

I think I'm going to faint, but instead
stroll across, force an introduction.
She has strong hair, a usurper's handshake
and 'nice to meet you' she says, dabbing
the corners of her mouth.

I retreat with grace to where D and ex-H
can report on how ex-L and companion are arranging
their faces. I play with my spaghetti marinara,
concentrate on other fish. I am
wonderfully serene.

The man at the next table tells us he's
my ex-husband's ex-lover's ex-husband, Neil.
Over an espresso, he announces
he's rethought his fellow-travelling
with feminism. He seems to want applause.

My ex knows Neil's ex- and current wives
are planning to move in together.

I know my ex-lover's woman is getting
in touch with her Inner Child and is giving
ex-L a hard time. Though not hard enough.

My daughter knows everything.

WHERE ARE YOU?

In this garden, after a day of rain,
a blackbird is making soundings,
flinging his counter-tenor line
into blue air, to where
an answering cadenza shows
the shape and depth of his own solitude.

Born in South London, inheritors
of brick, smoke, slate, tarmac,
uneasy with pastoral
as hill-billies with high-rise,
my parents called each other
in blackbird language:
my father's interrogative whistle
—'where are you?'
my mother's note, swooping, dutiful
—'here I am'.

There must have slid into the silences
the other questions,
blind, voiceless worms whose weight
cluttered his tongue;
questions I hear as, half a lifetime on,
I eavesdrop on blackbirds.

MOMENT

Driving to meet you,
a scent surfacing
in the mind
—a random pulse
leaping the wrong synapse—

the trace
of another life
—gunpowder, was it,
or hot vanilla—

overturned the furniture
of love,
as a shaft of air
from an opened window
enlarges an over-heated room.

THE PRESENT

Why, when I have ended it already
(seeing the exact shape of your cruelty)
should I dream this:

                    Standing in your house,
pinned by the weight of your coming marriage,
I'm power-dressed, try hard to envisage
a double funeral.

                    The room is crammed
with jewelled ornaments. 'Tell me', you say,
'which one should I give her on The Day,
the most exquisite thing?'

                    Objects pass
between us like formal kisses. To stifle tears,
and to advise at all, I have to lose
myself behind her eyes
                    —or take her in;
so when you choose a flower of chrysoprase
and garnets, smooth, translucent, and it lies
quivering in my palm,
                    I know: this is
the wonderful, the only present. 'Oh,
I would love this!' You take it from me. 'You
might,' you say,

                    'but, for her, it isn't
dangerous enough'—and then I see
the jewel is neat, ordinary, and I
am double-locked in my own skin

                    —but then,
knowing all this already, why that dream?
Unless there's something else I haven't seen.

ONE

Those are the worst times.
Not the cold
ghost at my back
when I turn in the small hours;

not the fury with myself
and you when the screwdriver's
wrong and my hand stumbles
and bleeds;

but after I've halved
the last piece of walnut cake,
or marked in the paper
something that would amuse you

or, at the sales, dithered
over whether you'd like
the check or the blue best
—the small ways

I assume you—then
the Oh, like an uppercut.
And look
I'm talking to you.

DIABOLUS IN MUSICA

Yours must be the coldest house in the city,
each window, veiled by swags of bleached cotton,
facing degrees of north. Too many times
I've waited for you in this house without odours,
fidgeted in matt rooms, naming the shades
—*anaemia*, *dandruff*, *asbestos*, *eau de mort.*

Now, vividly made up, dressed like a dancer,
I spread across your sofa drifts of magenta
and poppy print; but, pressing at the window,
the blotting-paper sky soaks up my brilliance.
A book? A drink? Cold has me by the throat.
I lift the lid of your white piano.

Tri-tone. The harsh sweetness shakes my fingers,
turns them to a devil's paint-brushes
which, pointed at the walls, begin to lift
the augmented fourth into spikes of colour,
wind-driven flame, the frenzy of flood water
coalescing into rafts of bubbles;

ground and sky, simultaneous—the cosmos
seen through a fish's eye. From widening cracks
insects, emerging as from chrysalids,
are starting to migrate from room to room.

It's enough. You'll make of it what you can.
I'm leaving your key, closing the door behind me.

Medieval composers found the tri-tone, or augmented fourth, an unpleasant interval, and called it 'diabolus in musica'.

OUR PEACOCK

He was a gloss on that English garden of roses,
banks of blowsy peonies, clipped box.
Ours, because we were his only audience,
and one of us, at least, wanting him to be
an oracle, to fill the sad silence between us
with a fanful of gorgeous air, a sign
richer than the sun's feeble water-dance.

Chivvied by a dozen bloomered bantams,
he dragged his train sulkily in the dust,
a legendary actor in a fit of temperament;
but turned then and, with a shiver of quills,
displayed his gifts to us, no holding back;
Platonic peacock strutting his stuff, all symmetry,
brilliancies, rank upon rank of exquisite eyes.

It wasn't nothing, that we were sharing this
in an English garden smelling of lavender.
But his cold glance told me there's no beauty,
anywhere, to set against old failures of love.
As we left, he lifted himself high into a tree,
and cried out; his voice, broken glass
tearing the heart out of the afternoon.

II

Striking Distance

OUT OF REACH

Raja . . . puja . . .
a Tamil song,
catchy in a Madras taxi,
seemed, in the space between near-misses,
to connect the driver's torn shirt,
the bullet-proof vest of the State governor,
quilts fitting the humped backs of cattle.

It's gone
leaving only its impression.
Asking in the music shop
I offer *Raja . . . puja . . .*
as a baby mouths urgent syllables
at kindly adults
deaf to the most important thing.

CROSSING THE BORDER

This cake I'm making—
I'd rather do almost anything else
but I need a place for these ingredients.

Elsewhere, another woman risks
a shell-gashed balcony to light a fire.
She guides a baby's thumb into its mouth.

My cake is made of dry and wet elements.
The god of the fleeting moment
blesses them into something new.

She has boiled pasta, a handful
for her family of four; a smear
of mustard. They call it soup.

This cake is made of such plenty
yet it won't rise.
I mean it as an offering

but how can it fit into a time
of bread in the wrong places,
of no—no more—nothing?

If I could, I'd walk with it
across the map of Europe,
over bland pastels, wavering boundaries,

to where she's silent as a man says,
I won't die of death, but of
Love For My City.

*

Children come like sparrows to my table
flight upon flight; cold
fingers grasp the hard edge,
nails scrabbling for grains of salt.

I eat my warm, rich food. Every day
they have a more migrant look.
Above them, the funeral bird
strops a complacent beak.

Sanity would turn time
back on itself, reel the children in,
to stack them
in vast ovarian warehouses, sleeping.

Let their next life be as meadow larks
—their high, clean thatness;
dying just deaths
unfreighted by love, pride, consequence.

STRIKING DISTANCE

Was there one moment when the woman
who's always lived next door turned stranger
to you? In a time of fearful weather
did the way she laughed, or shook out her mats
make you suddenly feel as though
she'd been nursing a dark side to her difference
and bring that word, in a bitter rush
to the back of the throat—*Croat / Muslim /*
Serb—the name, barbed, ripping
its neat solution through common ground?

Or has she acquired an alien patina
day by uneasy day, unnoticed
as fall-out from a remote explosion?
So you don't know quite when you came to think
the way she sits, or ties her scarf,
is just like a Muslim / Serb / Croat;
and she uses their word for water-melon
as usual, but now it's an irritant
you mimic to ugliness in your head,
surprising yourself in a savage pleasure.

Do you sometimes think, she could be you,
the woman who's trying to be invisible?
Do you have to betray those old complicities
—money worries, sick children, men?
Would an open door be too much pain
if the larger bravery is beyond you
(you can't afford the kind of recklessness
that would take, any more than she could);
while your husband is saying you don't understand
those people / Serbs / Muslims / Croats?

One morning, will you ignore her greeting
and think you see a strange twist to her smile
—for how could she not, then, be strange to herself
(this woman who lives nine inches away)
in the inner place where she'd felt she belonged,
which, now, she'll return to obsessively
as a tongue tries to limit a secret sore?
And as they drive her away, will her face
be unfamiliar, her voice, bearable:
a woman crying, from a long way off?

THANATOS

I was small again, pitting my mute will
against the clash of parents who held
my universe between them.

This time, though, the whole sky
was a battlefield for kamikaze planets
careening in a wide lens—bodies

sucking substance from each other
like punished souls caught, for want of pity,
in a timeless passion for each other's harm.

And Earth was part of it, with Mars,
its swelling shadow, our prognosis.
Half awake, I felt the world's oceans

pitching and slopping in the huge bowl
of my arms. Around me lay oblivious cities
and I was straining to hold steady, steady.

THERE WILL COME A TIME

from Marina Tsvetaeva

There will come a time, lovely creature,
when I shall be for you—a distant trace
lost in the blue pools of your memory.
You will forget my aquiline profile,
forehead wrapped in a halo of cigarette smoke,
the perpetual laugh I use to hoodwink people,
the hundred silver rings on my tireless hand,
this attic crow's-nest, the sublime confusion
of my papers.
                    You will not remember
how, in a terrible year, raised up by Troubles,
you were little.   And I was young.

(November, 1919)

ADVENT IN BRATISLAVA 1992

for Michael Rustin

Ten days to New Year, the fog's cold comfort
paints the square non-committal.
This passes for a market, ramshackle piles
of moist acrylic sweaters, glass baubles.
Children, like children anywhere,
suck chemical lollipops, staring
at the fog-wrapped, festive tree.

To us, from a more electric city,
this is a place in hibernation,
smatterings of light snuffed easily,
shops, cafés, turned inwards, as if refusing
to whore for passing trade, to get the hang
of the bland competence that's ordinary
up river, and points west.

We're plump with bright ideas, but this
is not the season for practicalities.
In the wedding cake Filharmonia,
fin de siècle bourgeois look-alikes,
treading the frayed red carpet, stroll
through the interval; their apparatchik past
tucked under their tongues, dissolving slowly.

On 1 January 1993, Czechoslovakia underwent a 'velvet divorce' to form separate Czech and Slovak republics.

THE TRIAL OF LYMAN ATKINS

Where he lives, they get along without them,
words. Him and his brothers. Just soft names
to coax the heifers out frosty mornings;
gabble, *gabble*, laughter teasing the turkeys;
once a month, greeting folks in the farmstore.

So he opens his empty mouth, and the words
cluster right where the lawyer puts them
I leaned on her face to stop her screaming
—like singing out *Amen*! to please the pastor,
like saying *I'm shit* to escape from the big guys
(words being light and slack as a string bag
shaped to whatever you want them to hold).

Or, being rare and rarely his own,
is it rather that words, once he hears himself speak them,
come to seem weighty and full of the truth?
Does he get to believe, for as long as he's talking,
that a girl screamed and he was the one who stopped her?
Does he imagine the press of her cheekbones?

Of course, in the nature of things, there's no knowing.
What's clear is he only half understands
—though he's learned the trick of seeming to, just by
repeating the words right after a person:
Yes, I waived my rights, and he's grinning
as if he's seeing himself in a motorcade,
while the jurors look grave, scribble on their note-pads
the words that dance round his head like horse-flies.

AMERICA

for Eva Hoffman

That was a haloed sound
the soft, crisp grace of it
held on the lips,
unfolding into space—
America!

We embarked in new clothes,
the plainest. We'd burned
the old, along with our sins,
all complicated, sad relations,
all slant, avaricious things.

Spring, and a following wind.
We almost rejoiced in spewing up
our very linings, as the gale
big-bellied the mainsail,
caked our hair white.

Our thoughts flew forwards,
singing. Had I understood
there is no innocence
like that of a long journey,
I might have dreaded landfall.

America—raccoon, cardinals,
new colours in the earth.
We trekked through summer,
guided to the place
we called Redemption.

All our wants were wants
for all of us. We were re-made
by daily quartering of bread;
scars, callouses in common.
We were perfected. Prospered . . .

Hard to say in whose bones
the seed of Self had lain like
tubercle. Slowly, we've sickened,
and with diseases one can't name
except by absence.

Our lips are flaking
as we creep into familiar rags
we don't acknowledge. We've become
all closed hands, faces, doors.
I can't live with no America.

SISTER SHIP

On 6 March 1987, the passenger ferry, *Herald of Free Enterprise*, capsized outside Zeebrugge, with the loss of 193 lives.

1 *Relatives*

More than a month
they've swallowed strange food,
found patience, grateful phrases,
while gales poured havoc over her.

The town's familiar as a lifer's cell.
At night, they lie counting traffic lights,
brands of beer—unconnected things
to keep them even.

Not much to say;
just waiting, walking streets,
the dock, staring out there,
trying to, trying not to picture him.

Waiting for cranes to raise her,
for the sea to give him back
so they can all go home,
so they can sleep.

2 *Divers*

She's come to term,
a monstrous animal
slouched onto one side
for the delivery.

The sea sulks, stands off.
Tugs bob, useless,
while they, like puny midwives,
delve in her cluttered entrails

half-blind, grasping
a thigh, a shoe,
an arm stiff round a pillar,
easing it free,

becoming
instant mortuary professionals,
prepared for the look
of face after face, defaced.

3 *Afterwards*

The duchess came,
gave him a toy,
said, *Brave little boy*.

The prince came,
admired her curls,
said, *Brave little girl*.

The Minister came,
shook his hand,
talked of England.

The pop star came
with a TV van,
said, *Terrible, man!*

Reporters came,
said, *Tell about the horror.*
Look at the camera.

The ambassador came,
the nun and the police,
the doctor and the priest.

But Daddy didn't come,
Mummy didn't come,
his wife didn't come,

her husband didn't come,
their daughter didn't come,
their sons didn't come,

and night has come
like a black stone
in the throat.

4 *Sister Ship*

Months afterwards
we travel in the sister ship,
our luggage stuffed with images
made toy-like by the geometry
of page or screen.

What can we do with them?
We're bit players in a re-take
with an uneventful ending
turning the one before to history;

though you can sense her weight,
her three and more dimensions
as she slides seaward, slowly turns.

5 *Not us*

That evening,
when the shutting of bow doors
was left to enterprise
and enterprise was sleeping,
would it have been like this:

you and I
out on A deck, watching
receding lights
spangle the ink black,
when suddenly

the sky tips
—funny at first—
pressing us on the rail.
I have inflatable armbands
fit them on you

—or do I make
a bowline from my coat-belt—
either way,
as the surface rushes up
we grasp hands, jump, swim.

Or did it list to starboard?
We're thrown backwards
against an iron stair
we cling to, wedge ourselves
waiting for rescue.

Would it have been
like that? You and I
survivors? Always a way
for those who keep their heads,
those with enterprise?

6 *Cafeteria*

This ship is full of empty passengers,
a cargo of consumers. One could think
we're working our passage with our teeth,
biting, chewing, swallowing our way
from shore to shore.

                    The cafeteria
fills up at once, a patient, silent queue
intent on breakfast: piles of sausages,
tomatoes, baked beans, bacon, eggs, fried bread
and toast and marmalade;

as though we've been
invaded by those taken by the sea,
ravenous spirits thrown up in the wake.
We glut ourselves to fill a double void
with obvious comforts, never quite enough.

7 *Anniversary*

They look pinched by a wind
stronger than then
and colder.

They're quiet as they queue
to scatter wreaths and posies
on the water,

shifty sea
that swallowed flesh
as if entitled.

So much salt
our cheeks are stiff with it.

So much salt
our eyes are glazed with it.

The wind spits back
their tears for kin
man-slaughtered,

not mattering enough;
and no one saying sorry.
They're here as witnesses

for those more than fractions
of the one-nine-three,
the paper dead.

So much salt
our stomachs sicken of it.

So much salt
our tongues are raw with it.

They twist their grief into
flowers, cast them adrift.
Anger, more difficult to place.

THIS ENGLAND

'I'm English' meant a welcome almost
everywhere; a safety-net woven of acronyms.
The best courts in the world,
live and let live, Jerusalem

fitted inside the cheek
as though that kind of Englishness
came with the water flowing
untainted into every home.

People like us never swallowed Glory;
we scorned the version where
dam busters posed on tiger rugs
behind a brute stockade of names for foreigners.

But Hope was part of it—stuffed envelopes,
instant solidarity, dawn pickets,
until our retreat to parenthood, DIY;
our bit—our neo-Marxist idiom.

Progress, the word, was out,
but we pick 'n mixed our England from the shelf
of attainable ideals. English meant
free to speak meant someone listened,

didn't they? But this is our back yard, this
cardboard habitat, shuffle of empty
Quality Street wrappers, this place where
climbers smother honesty and love lies bleeding.

We lost the England we assumed was everyone's.
Our heads were buried in each other's books
while asset-strippers grabbed the paving stones
and the bus moved off to Shopping City.

Our small, fastidious, disowning cries
are snaffled by the wind. It takes
more than a large and sorrowful vocabulary
to tangle with the warp of England.

TURNING POINT

You're standing in the supermarket queue
dreaming of difference;
or on a foreign street, you're weak
with possibilities there's no language for.
And you could walk
round the corner into that other place

but for the hours looping on themselves,
ordinary as knitting
you're too meshed in to see straight.
Each morning dazzles you with a tranche
of choices, distracting
from the sly way the needles skewer you.

Bad faith isn't high dramatic acts
—the Judas kiss, embezzlement—
it's allowing each innocuous stitch
to shift you from day to night
and back to day, no further on,
a little fitter for the shapeless droop

that lengthens to the final
casting off.
Out of what despair, or resolution,
might you refuse? And what
would guide you through
the chaos of your fraying ends?

CUTTING LOOSE

At one time
at moments when the walls burst through the wallpaper
he could pause long enough
to slam the door behind him as he fled
into the hard-boiled street;
so that coming home, hours later, all was intact
—walls, clock, kettle
subdued to simple ticks and shrieks and silences.

Now the door stands gaping
as he hurtles out, in sight of the street's main chancers
and he knows his room will be
picked clean and pissed on (though not with particular malice).
But it seems immaterial,
a room in a film—and the thought is a sudden joy
and a clarification:
instead of a loop, his route will be straight forward.

His feet make jet-streams.
He rides in the bowl of his pelvis like a millionaire
and he sheds desire
for the squared-up scraps of lives in the lighted windows
as his leg-springs carry him
through streets speckled with the whites of eyes,
past Coke can and condom,
the newsprint beds of the rootless, birthday-less,

to the waste-ground
where gaunt backs of warehouses stand, guarded
by hogweed and sycamore.
And though he knows he's matter out of place,
a kind of dirt,
he's a giant under the sky, chewing nettles
like cheekfuls of bliss;
for now, breathing easily, at home.

TUESDAY AT THE OFFICE

A human fly has climbed the pylon.
Stripped, but for shoes,
('I hope he gets sunburn,' Sandra says)
he's a cut-out against postcard blue.
We can't hear him, only see mimed anger
as he shouts down to firemen
perched on puny ladders, offering food.
('Waste of public money,' Sandra says)

It's like golf on TV.
We chat, type, photo-copy,
looking up between jobs
at our man, monkeying about.
If he's going to jump, we want to see,
be made solemn, half hoping
it'll make sense of something.
('I know it's a cry for help, but,' Sandra says)

At some point, he's gone.
That night, we watch Thames News
to prove it happened, and we were there.
But there's no mention,
only traffic and a murder,
so he must have climbed quietly
back into his ordinary life.
('I knew he'd let us down,' says Sandra)

FATIMA

Class, this is Fatima
all the way from—
who can spell Bosnia for me?

I know if she could speak
English, she would tell us
what a lucky girl she feels
to be here in Bromley—THIS IS BROMLEY—
while all her friends
had to stay behind in—
who can spell Sarajevo for me?

This morning we are going to carry on
with our Nativity Play For Today.
Fatima has lovely blonde hair—HAIR—
so she is going to play the Virgin Mary;
then she won't have to say anything.
No sulking, Lisa; you can be
the landlady. She's got a nice rude speech
and a shiny handbag.

Alex is Joseph; you other boys
are soldiers. But remember
you're not to get carried away
killing the babies. This is acting.

Fatima, sit here, dear;
this is your baby—BABY.
Joseph, put your hand on her shoulder.
Now, angel chorus, let's have the first verse
of 'Hope for the world, peace evermore.'
Herod, stop fidgeting with your kalashnikov.
Fatima, why are you crying?

BLURRED VISION

Out in the fast lane, life mimics
going places; no witnesses

to hands slithering on the wheel.
Tear vapour fogs the windscreen as,

on motorways to anywhere, men
cry in their private grief containers.

People they pass might be scandalized
by wide-open mouths boozily singing

but it's desperation unrefined
cracks such large holes in rigid faces.

Maybe they'd want to find themselves
in the hazy place before they were big boys;

but what they see is all one way,
mile upon mile of hard shoulder,

as they're driving, driving much too fast
to notice exits. Alone and dangerous.

LIFE IN TALL HOUSES

So many years of the tall, smart
Happy Family museums
insulting us by their indifference to blinds;

blazing rooms, boasting
amply laid tables, modish clutter,
children playing chamber music;

bait, perhaps.

Years of impregnable locks
until we came to imagine more intensely
those hugs, those conservatory flowers;

and the tall houses
cracked open like pomegranates
under the arithmetic of our desire;

a bit too easily.

The people sprang from their beds
with a curdled look, as though we
were what they'd always dreaded and needed.

The light inside the tall houses seems
misplaced, furniture paper-frail,
jasmine bent on dying.

We are left with a fistful of flies
and the thought of how the happy families
scattered into the city,

singing, or something like it.

WRONG-FOOTED

You're going down a long escalator
into the limbo of the Northern Line,
and despite all gratefully perceived distractions
—a fat balloon-seller gliding upwards,
Kew Gardens poster by a friend of yours—
you can see what's waiting at the bottom is
the subject of an undergraduate essay
you'd get a C for: weigh the respective claims
on the solitary coin that's in your pocket
of the flautist travestying *Summer Time*
and that grey bundle merging with the wall.
Giving to a busker is like shopping
(you're comfortable with that) paying for
a musical snack, instant insouciance;
it's something for something. She looks like your niece.
You don't want the despair of speechless begging
to chill your morning; that belongs outside
the world you think you live in, where one makes
an effort. So that when his body's angles
say life's shit-dark, and some twinge in your bones
says yes, you tell yourself that what you give
someone like that can never be enough,
or the right thing, that only love would do,
and what you've got is one coin in your pocket.
The unsolicited, junk notes sing
put it here . . . *the livin' is easy*. But you're thinking
maybe you shouldn't get by with a neat transaction;
this woman has her flute, and maybe smiles
into the mirror, liking the shape of her chin.
You want to turn and climb towards the light
but you know you'd only stay in the same place
—the thought from which the escalator trips you,
the essay in your head still at page one.

OURSTORY

Let us now praise women
with feet glass slippers wouldn't fit;

not the patient, nor even embittered
ones who kept their place,

but awkward women, tenacious with truth,
whose elbows disposed of the impossible;

who split seams, who wouldn't wait,
take no, take sedatives;

who sang their own songs, went uninsured,
knew best what they were missing.

Our misfit foremothers are joining forces
underground, their dusts mingling

breast-bone with scapula, forehead
with forehead. Their steady mass

bursts locks; lends a springing foot
to our vaulting into enormous rooms.

THE SMELL OF SWEAT

Sweat is our signature on air:
grapefruit, onions, Glenmorangie.

It is the first date,
the first exam, all firsts;

climb on the cliff path, straining
to work off a terminal prognosis;

rugby hugger-mugger, job well done,
the body throwing out exuberant salts;

nightmare-time—debt, the law,
ex-husband at the door;

one of lust's slippery bouquet
of juices; must of the prison cell.

Its traces swirl around us;
we draw daily breath

from this promiscuous reserve
of extreme moments, not knowing whose.

OXFORD POETS

Fleur Adcock
Moniza Alvi
Kamau Brathwaite
Joseph Brodsky
Basil Bunting
Daniela Crăsnaru
W. H. Davies
Michael Donaghy
Keith Douglas
D. J. Enright
Roy Fisher
Ida Affleck Graves
Ivor Gurney
David Harsent
Gwen Harwood
Anthony Hecht
Zbigniew Herbert
Thomas Kinsella
Brad Leithauser
Derek Mahon
Jamie McKendrick
Sean O'Brien
Peter Porter
Craig Raine
Zsuzsa Rakovszky
Henry Reed
Christopher Reid
Stephen Romer
Carole Satyamurti
Peter Scupham
Jo Shapcott
Penelope Shuttle
Anne Stevenson
George Szirtes
Grete Tartler
Edward Thomas
Charles Tomlinson
Marina Tsvetaeva
Chris Wallace-Crabbe
Hugo Williams